Bibliographic information published by the German National Library:

The German National Library lists this publication in the National Bibliography; detailed bibliographic data are available on the Internet at http://dnb.dnb.de .

Imprint:

Copyright © 2017 GRIN Verlag, Open Publishing GmbH
Print and binding: Books on Demand GmbH, Norderstedt Germany
ISBN: 9783668383432

This book at GRIN:

http://www.grin.com/en/e-book/351564/molecular-phylogenetic-studies-of-pumpkin-cucurbita-argyrosperma-and

Prem Jose Vazhacharickal, Sajeshkumar N.K, Jiby John Mathew, Anjana R, Dona Ann Johns

Molecular phylogenetic studies of pumpkin (Cucurbita argyrosperma) and winter melon (Benincasa hispida)

Family comparison using rbcL sequence analysis

GRIN Publishing

Molecular phylogentic studies of pumpkin (*Cucurbita argyrosperma*) and winter melon (*Benincasa hispida*): family comparison using rbcL sequence analysis

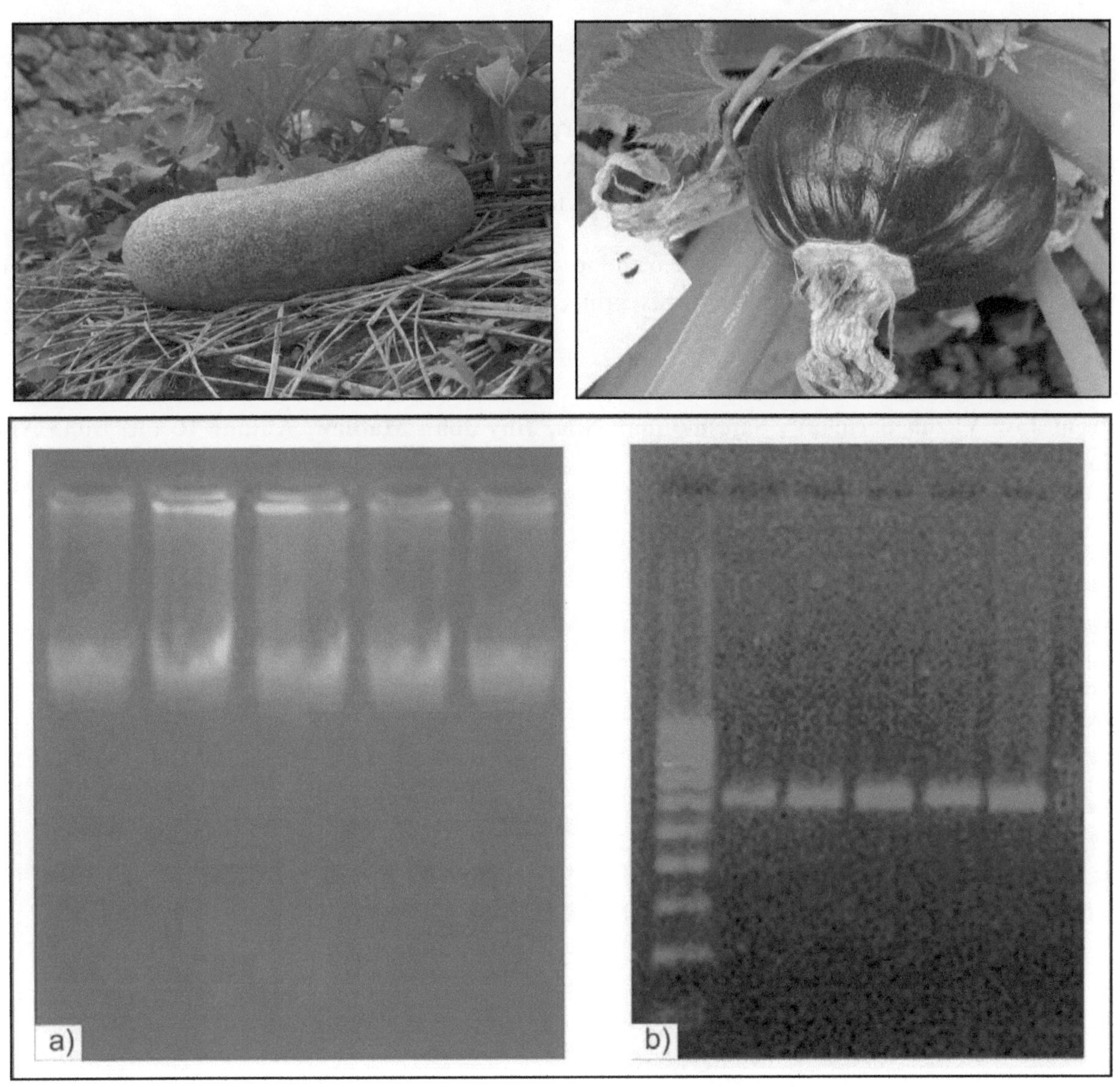

Prem Jose Vazhacharickal, Sajeshkumar N.K, Jiby John Mathew, Anjana R and Dona Ann Johns

Acknowledgements

Firstly we thank **God Almighty** whose blessing were always with us and helped us to complete this project work successfully.

We wish to thank our beloved Manager **Rev. Fr. Dr. George Njarakunnel,** Respected Principal **Dr. Joseph V.J,** Vice Principal **Fr. Joseph Allencheril,** Bursar **Shaji Augustine** and the Management for providing all the necessary facilities in carrying out the study. We express our sincere thanks to **Mr. Binoy A Mulanthra** (lab in charge, Department of Biotechnology) for the support. This research work will not be possible with the co-operation of many farmers.

We are gratefully indebted to our teachers, parents, siblings and friends who were there always for helping us in this project.

Prem Jose Vazhacharickal*, Sajeshkumar N.K, Jiby John Mathew, Anjana R and Dona Ann Johns

*Address for correspondence
Assistant Professor
Department of Biotechnology
Mar Augusthinose College
Ramapuram-686576
Kerala, India
premjosev@gmail.com

Table of contents

Table of figures

Table of tables

List of abbreviations

BLAST	: Basic local alignment search tool
CTAB	: Cetyl trimethylammonium bromide
FASTA	: Fast All
HTML	: Hyper text markup language
$MgCl_2$	: Magnesium chloride
NJ	: Neighbour joining
PCR	: Polymerase chain reaction
RAPD	: Random amplified polymorphic DNA
rbcL	: Ribulose bisphosphate carboxylase large chain
Rubisco	: Ribulose-1, 5-bisphosphate carboxylase oxygenase
SPSS	: Statistical package for social sciences
Tag	: *Thermus aquaticus*

Molecular phylogenetic studies of pumpkin (*Cucurbita argyrosperma*) and Winter melon (*Benincasa hispida*): family comparison using rbcL sequence analysis

Prem Jose Vazhacharickal[1]*, Sajeshkumar N.K[1], Jiby John Mathew[1], Anjana R[1] and Dona Ann Johns[1]

[1]*Department of Biotechnology, Mar Augusthinose College, Ramapuram, Kerala, India*

** Corresponding author; premjosev@gmail.com*

Abstract

The Cucurbitaceae family is the one of the economically important group of plants in the tropics and subtropics. Molecular phylogenetic analysis were advanced after the introduction of molecular markers which give much precise results in analysis. Our current study based on the amplification of RuBisco enzyme using rbcL primer and subsequent validation using BLAST, FASTA and CLUSTAL-W in pumpkin and winter melon. The isolated and purified DNA samples were PCR amplified using rbcL primer and later sequenced using ABI Prism 377 DNA sequencer. Multiple sequence alignment algorithms and distance matrix were constructed using rbcL sequences in FASTA format were retrieved from GenBank. Phylogenetic tree was created using the distance based neighbour joining (NJ) and clustering algorithms method. Once divergences between all pairs of samples were determined, statistical cluster analysis and dendrogams examines the similarity among halotypes. Bootstrapping and jackknifing further increase the reliability estimates for the position of haplotypes within the evolutionary tree.

Keywords: Clustering algorithms; rbcL sequences; Multiple sequence alignment.

1. Introduction

Genetics, genomic and plant breeding has been considered as three overlapping and complementary disciplines for crop improvement. The Cucurbitaceae family has 118 genera and 825 species mainly distributed in the tropics and subtropics (Esteras et al., 2011; Jeffrey, 2005) which are considered to be the economically important plant families (Kocyan et al., 2007). The genus Cucurbita (2n = 2x = 40) has a great diversity in physiological and morphological characteristics (Robinson and Decker-Walters, 1997; Goldman, 2004; Gong et al., 2013). Studies were conducted to characterize the various tribes of Cucurbitaceae based on tendril branching (Jeffrey, 2005), pollen structure and seed coat characterization (Kocyan et al., 2007). With the evolution and advancement of molecular marker tools, The Cucurbitaceae has been selected as an ideal candidate for genetic scrutiny (Lebeda et al., 2007; Gong et al., 2013).

Cucurbitaceae (gourd family) is an excellent example of a plant family with many useful species of agricultural importance that are cultivated worldwide. The earliest record of human use of edible cucurbits comes from habitations older than 9000 years and certainly by 3000 B.C. In the new world, squashes and pumpkins are used as major food crop by native people. This is an important family consisting of approximately 125 genera and 960 species, mainly in regions tropical and subtropical. Cucurbita fruits have yellow or orange flesh, which is rich in carotenoids, the compounds humans need to make vitamin A and our visual pigment rhodopsin.

A taxonomic investigation of the cucurbits (family-Cucurbitaceae) growing throughout the northern parts of Bangladesh was carried out. A total of 24 species under 13 genera of the family Cucurbitaceae were collected. A brief taxonomic account of each species is given with current nomenclature, local name, English name and uses. About the species Benincasa hispida, Citrullus lanatus, Coccinia grandis, Cucumis sativus, C. melo, Cucurbita maxima, Cucurbita argyrosperma, C. moschata, C. pepo, Lagenaria siceraria, Luffa acutangula, L. cylindrica, Momordica charantia, M. cochinchinesis, Melothria maderaspatana, Trichosanthes anguina, T. dioica,T. cucumerina, T. bracteata were abundant and Diplocyclos palmatus, Cucumis callosus, Thladianthacordifolia, Trichosanthes chordata, Gymnopetalum cochinchinense were very rare.

Most of the research done using molecular markers usually geographic, country specific while some research were done across the globe for worldwide genetic variations (Gwanama et al., 2000; Paris et al., 2003; Wu et al., 2011; Gong et al., 2012). Investigations based on allozyme (Wilson, 1989), ribosomal intergenic spacer probes (Ruiz and Hemleben, 1991), chloroplast DNA (Wilson et al., 1992), and RAPDs (Baranek et al., 2000; Ferriol et al., 2003). Nuclear DNA markers are more reliable than organelle DNA due to the inheritance from both parents as well as meiotic recombination (Gong et al., 2013). Rubisco (Ribulose-1, 5-bisphosphate carboxylase oxygenase) is an enzyme found in plant chloroplast, has a vital function in Calvin cycle and glucose synthesis.

The rbcL gene and protein sequences have been used in addressing systematic questions among the few selected members of the family Cucurbitaceae has been investigated. In order to elucidate the systematic positions, a set of chloroplast-rbcL nucleotide sequences (from 42 taxa of 7 genera) and aminoacid sequences (from 52 taxa of 10 genera) were withdrawn from GenBank and GenPept databases, respectively. The evolutionary distance was inferred from these sequences by employing Bootstrap method of UPGMA (Unweighted Pair Group Method with Arithmetic Mean) and MP (Maximum Parsimony) using MEGA (Molecular Evolutionary Genetic Analysis) software. From the separate analysis produced almost similar although not identical results, no strongly supported incongruent results. The members of the genus Austrobryonia showed strictly monophyletic, Trichosanthes, Luffa, Momordica and Coccinia are found to be paraphyletic. But the members of the genus Cucumis are distributed throughout these hiraeoid clades, confirming the polyphyly of this large genus observed in both the family trees. From the results, it is also clear that, the chloroplast-rbcL gene and aminoacid sequences resolved the relationships, as well as provided a good indication of major supra-generic groupings among the selected members of the family Cucurbitaceae

Molecular phylogentic approach which belongs to molecular systematic, uses the differences in DNA sequences to determine the evolutionary relationship based on phylogenic tree. Comparison of homologous sequences for genes using sequence alignment techniques and data base search is a good indicator of degree of divergence during evolution. Given lacking information about molecular phylogeny of pumpkin and winter melon, our objectives were to (1) Amplify the genes producing the RuBisco enzyme using rbcL primer, (2) validation of the amplified genes using similarity searches including BLAST and FASTA and (3) interpretation of the software algorithms using CLUSTAL-W.

2. Materials and methods

2.1 Plant samples collection

Fresh plant materials of pumkin (*Cucurbita argyrosperma*) and winter melon (*Benincasa hispida*) was collected from a local farm in Melukavu (Figure 1). Fresh leaves samples were collected in pre-sterilized autoclavable polyethylene bags (Himedia Laboratories, Mumbai, India) and stored in insulated boxes with gel ice packs during transportation. The samples were stored at -20°C till further analysis. For the extraction of DNA, alkaline lysis method (Kidwell and Osborn, 1992) followed by modified C-TAB (Porebski et al., 1997) method. The quality of the isolated DNA was determined spectrophomertically using a double beam UV spectrophotometer (118, Systronics India Ltd, Ahmedabad, India) as well as 8% agarose gels, 1x TBE buffer in an agarose electrophoresis chamber (Geni Pvt, Bangalore, India).

The amplification of the extracted DNA were done using PCR machine (MJ Mini, BioRad, Gurgaon, India) according to the protocol of Zang and Renner (2003) using Taq Polymesase (Geni Pvt, Bangalore, India) and rbcL primer (Geni Pvt, Bangalore, India). The amplification performed in 25 µl of 25 µmol l^{-1} MgCl$_2$ solution, 2.5 µl 10 x buffer (Geni Pvt, Bangalore, India), 2 µl of a 2.5 µmol l-1 dNTP solution, 1 µl of rbcL primer at 10 pmol µl^{-1}, 1 unit (0.2 µl) of Taq Polymerase, and 1 µl of extracted DNA (Haas et al., 2003; Rychlik et al., 1990). The forward primer rbcL a F (5' ATGTCACCACAAACAGAGACTAAAGC 3') and reverse primer rbcL a R (5' GTAAATCAAGTCCACCACG 3').

After the PCR amplification, the PCR products were separated using agarose gel (1.5%) electrophoresis with a 100 base pair standard ladder DNA of 100 to 1000 BP (Geni Pvt, Bangalore, India). After agarose gel electrophoresis, the PCR products were trimmed from the gel, subsequently the gel was dissolved and later DNA was precipitated using 98% ethanol as precipitating agent (Meier et al., 1996; Erlich, 1989; Saiki, 1989). The purified DNA samples were later sequenced using ABI Prism 377 DNA sequencer (Applied Biosciences, NY, USA).

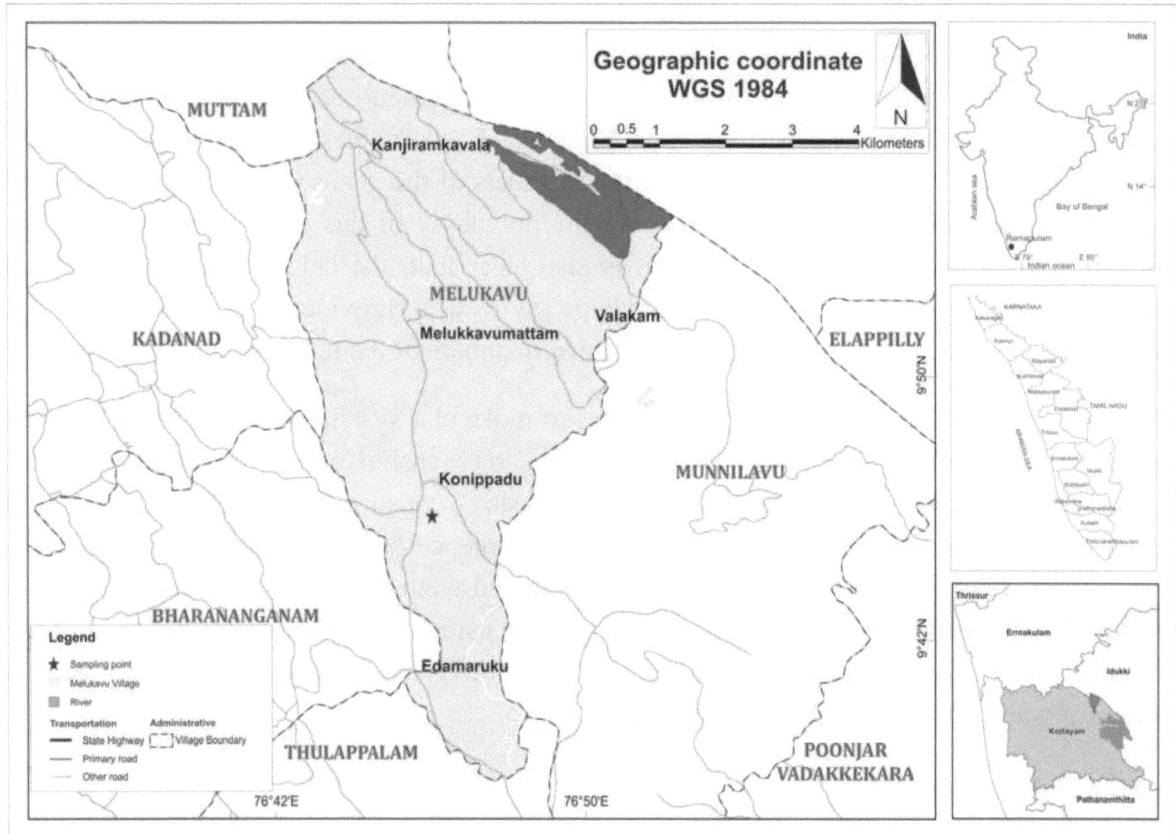

Figure 1. Map of the sample collection area represented by a star symbol.

Figure 2. a) Bitter gourd (*Momordica charantia*) fruits, b) Bitter gourd flowers, c) *Coccinia grandis* developing fruit, d) *Coccinia grandis* harvested mature fruits, e) winter melon (*Benincasa hispida*) fruit, f) *Cucurbita maxima* fruit. Photo courtesy: Wikipedia.

Table 1. Components of the PCR reaction mixture

Components	Volume/reaction
De-ionized water	16.5μl
Taq buffer (10X)	2.5 μl
Template DNA (20ng/μl)	2.0 μl
$MgCl_2$ (15Mm)	1.0 μl
dNTPs mix (10mMm each)	1.5 μl
Primer mix (10pm/μl)	1.0 μl
Forward primer	0.5μl
Reverse primer	0.5μl
Taq DNA polymerase (5U/μl)	0.5 μl
Final volume	25.0 μl

Table 2. PCR cycle for rbcL amplification

Cycle	Step	Process	Temperature	Time period
1	1	Initial denaturation	95°C	3 minutes
1	2	Denaturation	95°C	30 seconds
1	3	Annealing	55°C	30 seconds
1	4	Primer extension	72°C	30 seconds
2-32		Repeat steps 2 to 4 for 32 times		
33	5	Final elongation	72°C	10 minutes
		End or 4°C for ever		

2.2 Analysis of the sequences BLAST and Clustal-W

Basic Local Alignment Search Tool (BLAST) is used sensitive sequence similarity between sequences. The sample sequences in FASTA format were entered in field specified on BLAST algorithms home page. The BLAST output parameters were set to pre-determined settings and final output in Hyper Text Mark Up (HTML) format were used for further bioinformatics analysis.

Multiple sequence alignment programme Clustal-W 2.1 (University College of Dublin, Ireland). The process of Clustal multiple sequence analysis was advanced through pair wise alignment, construction of a distance matrix and display of base saturation in the matrix. The results were based on the pair wise analysis of 19 sequences using Maximum Composite Likelihood Method (Thompson et al., 1994; Chenna et al., 2003). Codon positions included were 1st + 2nd + 3rd + non coding sequences. Positions containing gaps and missing data were eliminated from the data set using complete deletion option to achieve high resolution phylogenetic tree construction.

In the case of bitter gourd, for precise results in the tree construction, 17 cucurbit samples and their respective rbcL sequences (GenBank) were included in the multiple sequence alignment algorithms along with *Momordica charantia*, *Cucurbita maxima*, *Coccinia grandis* and *Benincasa hispida*. For pumpkin, 15 cucurbit (other than *Cucurbita argyrosperma*) and an out-group *Solanum melongena* (brinjal) and their rbcL sequences in FASTA format were retrieved from GenBank. These were included in the multiple sequence alignment algorithms and in construction of distance matrix. The purpose of out-group is to increase the rate of interpretability of phylogenetic tree which shows maximum rate of evolutionary divergence.

Table 3. Sequences in FASTA format retrieved from GenBank for multiple sequence alignment algorithms and construction of distance matrix for winter melon.

No	Analysed sample	Retrieved sample species from GenBank
1	*Momordica charantia*	*Momordica foetida, Momordica cochinchinesis*
2	*Benincasa hispida*	*Benincasa fistulosus, Benincasa colocynthis*
3	*Cucurbita maxima*	*Cucurbita argyrosperma, Cucurbita palmate, Cucurbita foetidissima*
4	*Coccinia grandis*	*Coccinia quinqueloba, Coccinia adoensis, Coccinia mackenii*

Table 4. Sequences in FASTA format retrieved from GenBank for multiple sequence alignment algorithms and construction of distance matrix for pumpkin.

No	Analysed sample	Retrieved sample species from GenBank
1	*Momordica charantia*	*Momordica foetida, Momordica cochinchinesis*
2	*Benincasa hispida*	*Benincasa fistulosus, Benincasa colocynthis*
3	*Cucurbita maxima*	*Cucurbita argyrosperma, Cucurbita palmate, Cucurbita foetidissima*
4	*Coccinia grandis*	*Coccinia quinqueloba, Coccinia adoensis, Coccinia mackenii*
5		*Solanum melongena* (out-group)

From the preliminary results of multiple sequence alignment and distance matrix, non-aligned and mismatched sequences were removed for better reliability. Further this makes the understand ability of evolutionary convergence and divergences of the selected samples as well as with the other 11 common cucurbits included.

2.3 Construction of phylogenetic tree

The evolutionary relationships can be best studied using a phylogenetic tree and was created using the distance based neighbour joining (NJ) and clustering algorithms. Multifurcating phylogenetic tree was constructed based on creating a distance matrix starting from the alignment and pair wise distances calculated between sequences from the obtained matrix.

2.4 Statistical analysis

Descriptive statistics using SPSS 12.0 (SPSS Inc., Chicago, IL, USA) were conducted to summarize the data and graphs were generated using Sigma Plot 7 (Systat Software Inc., Chicago, IL, USA).

3. Results

The isolated DNA were checked spectrophotometrically and with a purity values of 1.67 and with a concentration of 4,395 µgml^{-1}. The agarose gel electrophoresis were used to separate the isolated as well as PCR amplified DNA samples. The sequences obtained after sequencing were subjected to BLAST similarity search and multiple sequence alignment using CLUSTAL-W.

The construction of divergence matrix and phylogenetic dendrograms revealed two main groups, one consisting of MC and other Momordica species while the other group consisted of three subgroups of CM, CG and BH having members of Cucurbita, Coccinia and Momordica respectively.

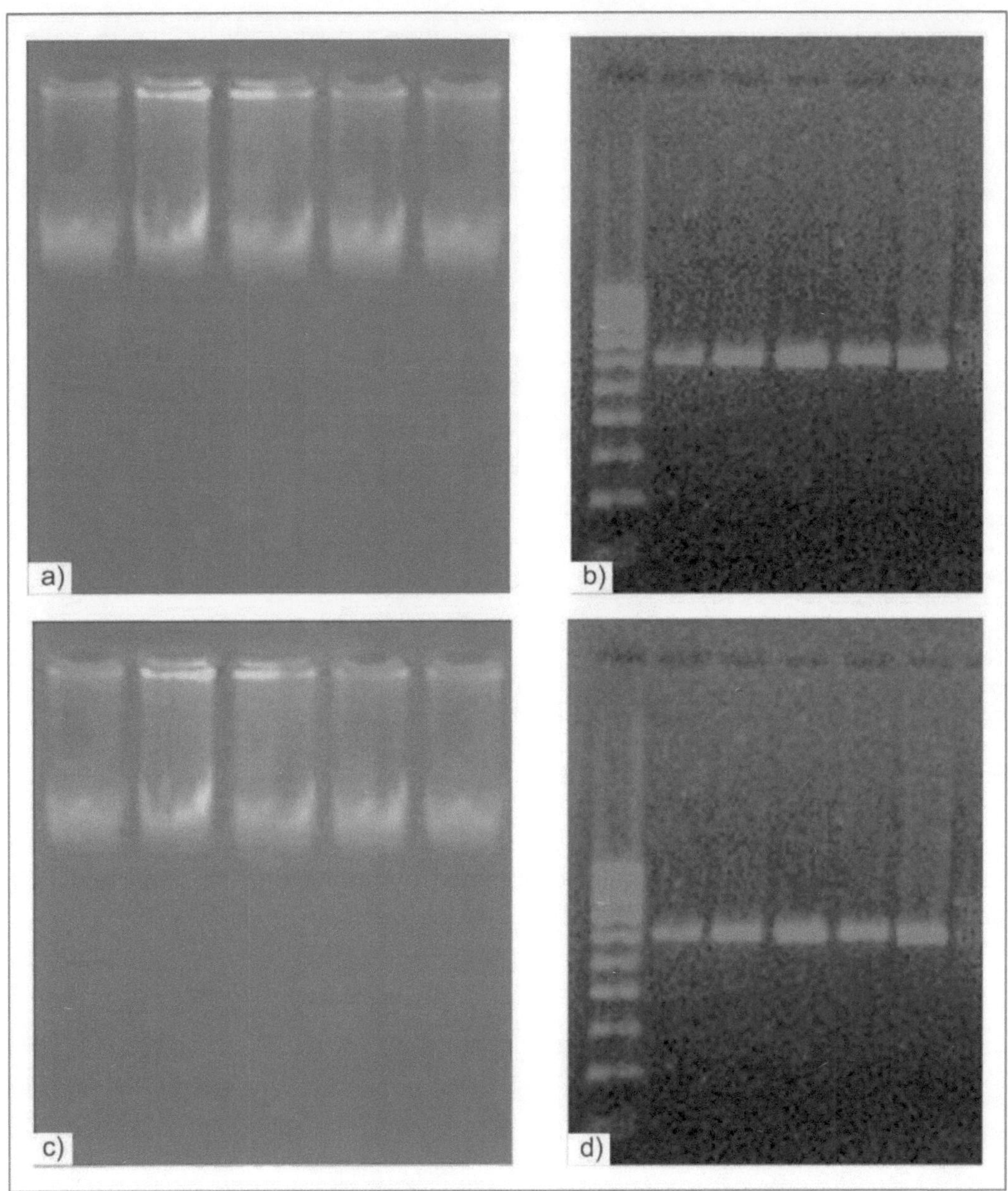

Figure 3. a) Agarose gel electrophoresis of the isolated DNA samples and b) PCR amplified DNA samples from winter melon. c) Agarose gel electrophoresis of the isolated DNA samples and d) PCR amplified DNA samples from pumpkin.

Accession	Description	Max score	Total score	Query coverage	E value	Max ident
HQ438613.1	Cucurbita argyrosperma voucher PI 163591 ribulose-1,5-bisphospha	830	830	100%	0.0	100%
JN796936.1	Cucurbita foetidissima ribulose-1,5-bisphosphate carboxylase/oxyge	824	824	100%	0.0	99%
HQ438634.1	Cucurbita palmata ribulose-1,5-bisphosphate carboxylase/oxygenas	824	824	100%	0.0	99%
HQ438632.1	Cucurbita foetidissima voucher PI 435094 ribulose-1,5-bisphosphate	824	824	100%	0.0	99%
HQ438627.1	Cucurbita maxima voucher PI 532715 ribulose-1,5-bisphosphate car	824	824	100%	0.0	99%
DQ535845.1	Sicana odorifera voucher Renner 2807 (M), cult. Munich BG from se	824	824	100%	0.0	99%
DQ535743.1	Calycophysum pedunculatum ribulose-1,5-bisphosphate carboxylase	824	824	100%	0.0	99%
L21938.1	Cucurbita pepo ribulose 1,5-bisphosphate carboxylase large subunit	824	824	100%	0.0	99%
HQ438631.1	Cucurbita ficifolia voucher PI 442180 ribulose-1,5-bisphosphate car	819	819	100%	0.0	99%
HQ438622.1	Cucurbita pepo subsp. fraterna voucher PI 614683 ribulose-1,5-bisp	819	819	100%	0.0	99%
DQ535844.1	Selysia prunifera voucher H. Hentrich FGIC60 (ULM) ribulose-1,5-bis	819	819	100%	0.0	99%
DQ535804.1	Cucurbita ficifolia voucher Renner et al. 2766 (M), cult. Mainz BG ril	819	819	100%	0.0	99%
DQ535790.1	Cionosicyos macranthus voucher E. Cruz, 6 May 2002, unvouchered	819	819	100%	0.0	99%
HM849900.1	Citrullus colocynthis ribulose-1,5-bisphosphate carboxylase/oxygen	813	813	100%	0.0	99%
HM849901.1	Citrullus lanatus ribulose-1,5-bisphosphate carboxylase/oxygenase	813	813	100%	0.0	99%
DQ535852.1	Tecunumania quetzalteca voucher L. D. Gomez 20988 (MO) ribulose	813	813	100%	0.0	99%
DQ535834.1	Penelopeia suburceolata voucher A. Veloz, A. Guerrero & N. Ramirez	813	813	100%	0.0	99%
DQ535791.1	Citrullus colocynthis voucher Renner et al. 2762 (M), cult. Mainz BG	813	813	100%	0.0	99%
DQ535785.1	Borneosicyos simplex voucher SAN (Postar et al.) 144251 (L) ribulo	813	813	100%	0.0	99%
DQ535778.1	Acanthosicyos horridus voucher E. van Jaarsveld s.n., cult. Kirstenl	813	813	100%	0.0	99%
DQ535769.1	Schizocarpum palmeri ribulose-1,5-bisphosphate carboxylase/oxyge	813	813	100%	0.0	99%
DQ535768.1	Schizocarpum filiforme ribulose-1,5-bisphosphate carboxylase/oxyg	813	813	100%	0.0	99%
DQ535745.1	Citrullus lanatus ribulose-1,5-bisphosphate carboxylase/oxygenase	813	813	100%	0.0	99%
AF008961.1	Abobra tenuifolia ribulose bisphosphate carboxylase/oxygenase larg	813	813	100%	0.0	99%
HM596956.1	Cucumis sp. IT-2010b ribulose-1-5-bisphosphate carboxylase/oxyg	808	808	100%	0.0	99%
HM596957.1	Cucumis sp. IT-2010b ribulose-1-5-bisphosphate carboxylase/oxyg	808	808	100%	0.0	99%
HM596954.1	Cucumis sp. A PMS-2010 ribulose-1-5-bisphosphate carboxylase/ox	808	808	100%	0.0	99%
EU541411.1	Zehneria anomala ribulose-1,5-bisphosphate carboxylase/oxygenase	808	808	100%	0.0	99%
DQ535837.1	Praecitrullus fistulosus voucher D. Decker-Walters 883 (FTG) ribulo	808	808	100%	0.0	99%

Figure 4. Blast analysis of the query sequence obtained by subjecting the amplified PCR products of pumpkin which are sequenced by ABI377 sequencer.

```
            ....|....10....|....20....|....30....|....40....|....50....|....60....|....70....|....80...
            CCGGGAGTTCCACCCGAGGAAGCAGGGGCCGCTGTAGCTGCTGAATCTTCTACTGGTACATGGACAACTGTGTGGACCGATGG
foetida         ..........................................................................
cochinchinensis .......................................................C..................
charantia       ..........................................................................
              .....................T....................................................
hispida1        .....................T....................................................
hispida2        .....................T....................................................
fistulosus      .....................T....................................................
colocynthis     .....................T....................................................
              ...............................G..........................................
argyrosperma    ...............................G..........................................
palmata         ..........................................................................
foetidissima    ..........................................................................
maxima          ..........................................................................
              .....................T....................................................
grandis         .....................T....................................................
quinqueloba     .....................T....................................................
adoensis        .....................T....................................................
mackenii        .....................T....................................................

            ....|....110...|....120...|....130...|....140...|....150...|....160...|....170...|....180
            GTTACAAAGGACGATGCTATGGCATCGAGCCTGTTCCTGGAGAAGAAAGTCAATTTATTGCTTATGTAGCTTATCCCCTAGAC
foetida         ..........................................................................
cochinchinensis .......................C..................................................
charantia       ..........................................................................
              ......................A...........................A.....A..................
hispida1        ......................A...........................A.....A..................
hispida2        ......................A...........................A.....A..................
fistulosus      ......................A...........................A.....A..................
colocynthis     ......................A...........................A.....A..................
              ......................A...........................A.....A..................
argyrosperma    ......................A...........................A.....A..................
palmata         ......................A...........................A.....A..................
foetidissima    ......................A...........................A.....A..................
```

Figure 5. Multiple sequence alignment of the 19 input sequences using MEGA4 software with sequences bootstrapped to 180 base pairs (winter melon).

	1	2	3	4	5	6	7	8	9	10	11	12	13	14	15	16	17	18	19
1. MC																			
2. M foetida	0.007																		
3. M cochinchinensis	0.016	0.018																	
4. M charantia	0.000	0.007	0.016																
5. BH	0.027	0.030	0.034	0.027															
6. B hispida1	0.027	0.030	0.034	0.027	0.000														
7. B hispida2	0.027	0.030	0.034	0.027	0.000	0.000													
8. P fistulosus	0.027	0.030	0.034	0.027	0.000	0.000	0.000												
9. C colocynthis	0.025	0.027	0.032	0.025	0.002	0.002	0.002	0.002											
10. CM	0.023	0.025	0.030	0.023	0.009	0.009	0.009	0.009	0.007										
11. C argyrosperma	0.023	0.025	0.030	0.023	0.009	0.009	0.009	0.009	0.007	0.000									
12. C palmata	0.020	0.023	0.027	0.020	0.007	0.007	0.007	0.007	0.004	0.002	0.002								
13. C foetidissima	0.020	0.023	0.027	0.020	0.007	0.007	0.007	0.007	0.004	0.002	0.002	0.000							
14. C maxima	0.020	0.023	0.027	0.020	0.007	0.007	0.007	0.007	0.004	0.002	0.002	0.000	0.000						
15. CG	0.027	0.034	0.039	0.027	0.009	0.009	0.009	0.009	0.007	0.013	0.013	0.011	0.011	0.011					
16. C.grandis	0.027	0.034	0.039	0.027	0.009	0.009	0.009	0.009	0.007	0.013	0.013	0.011	0.011	0.011	0.000				
17. C quinqueloba	0.025	0.032	0.037	0.025	0.007	0.007	0.007	0.007	0.004	0.011	0.011	0.009	0.009	0.009	0.002	0.002			
18. C adoensis	0.025	0.032	0.037	0.025	0.007	0.007	0.007	0.007	0.004	0.011	0.011	0.009	0.009	0.009	0.002	0.002	0.000		
19. C mackenii	0.025	0.032	0.037	0.025	0.007	0.007	0.007	0.007	0.004	0.011	0.011	0.009	0.009	0.009	0.002	0.002	0.000	0.000	

Figure 6. Estimates of evolutionary divergences between 19 input sequences using maximum composite likelihood method in MEGA4 software (winter melon).

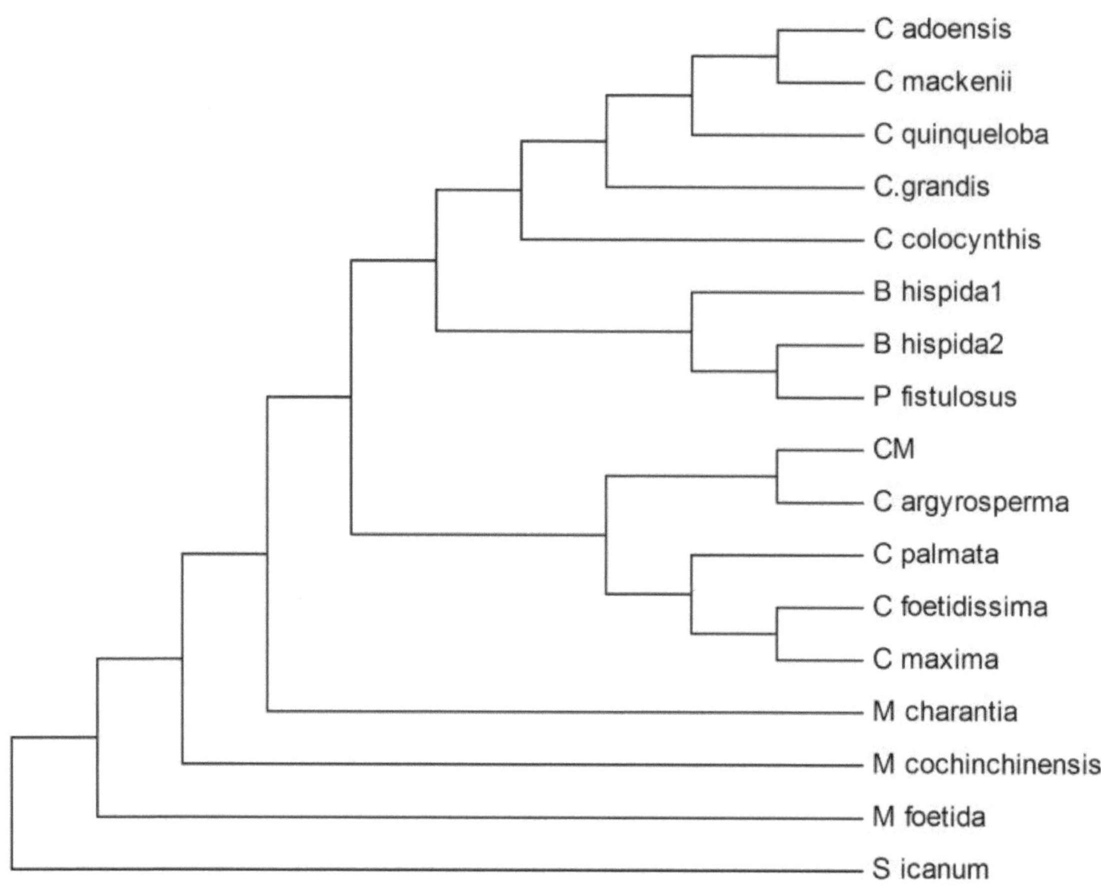

Figure 7. Evolutionary history of 19 taxa determined using neighbour-joining method (winter melon).

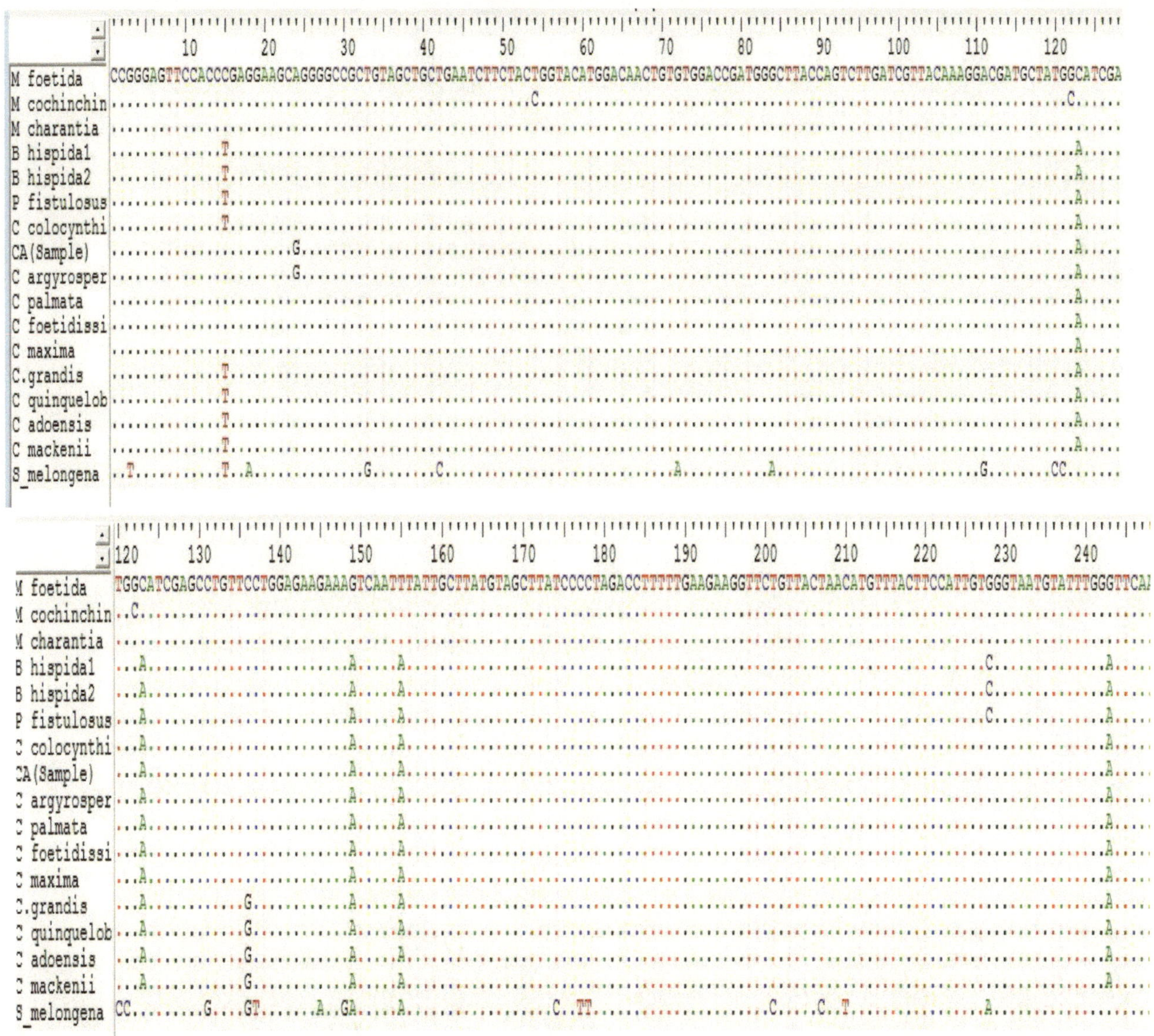

Figure 8. Multiple sequence alignment of the 16 input sequences and out-group using MEGA4 software with sequences bootstrapped to 240 base pairs (winter melon).

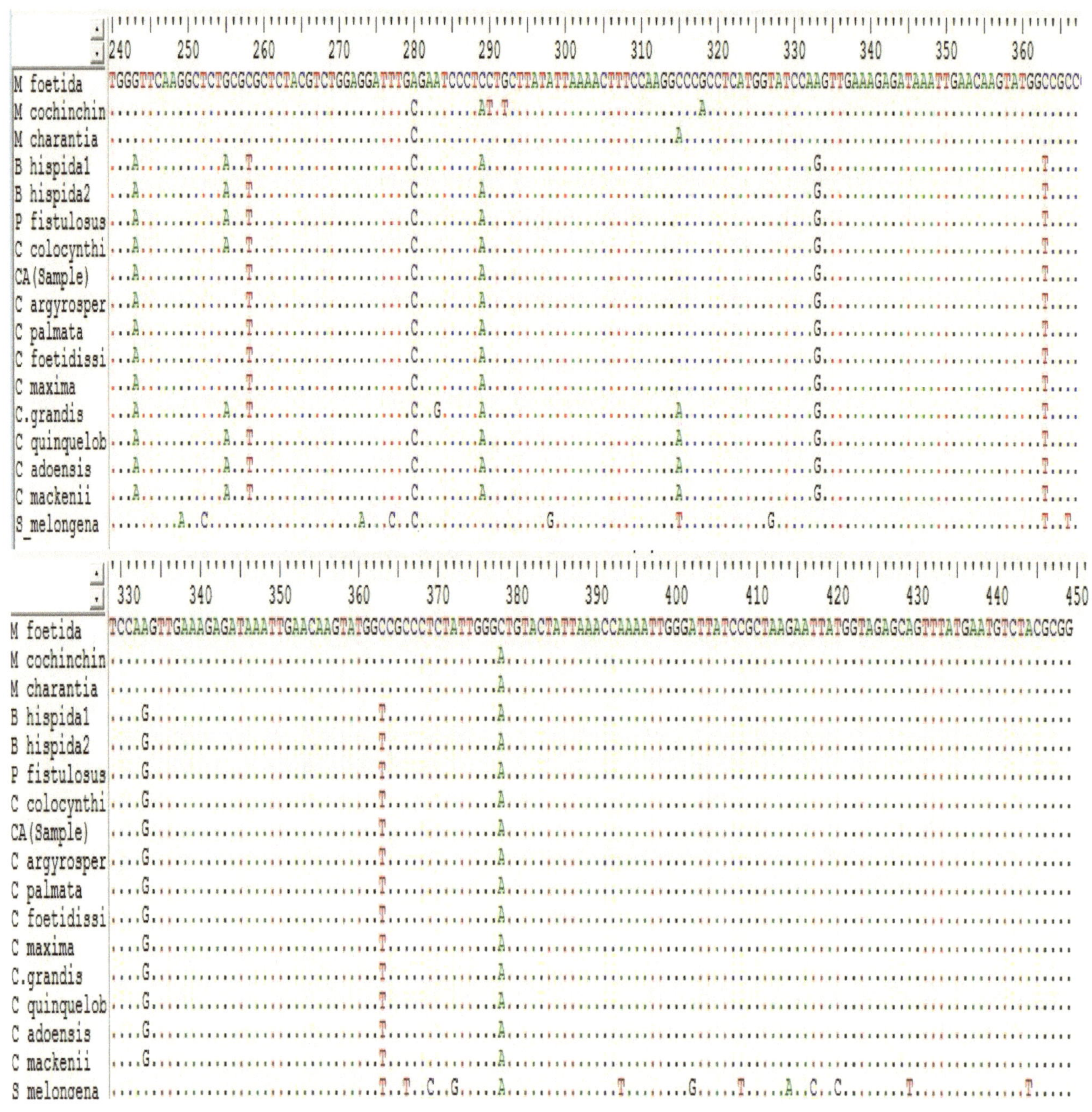

Figure 9. Multiple sequence alignment of the 16 input sequences and out-group using MEGA4 software with sequences bootstrapped to 450 base pairs (winter melon).

	1	2	3	4	5	6	7	8	9	10	11	12	13	14	15	16	17
1. M foetida																	
2. M cochinchinensis	0.018																
3. M charantia	0.007	0.016															
4. B hispida1	0.030	0.034	0.027														
5. B hispida2	0.030	0.034	0.027	0.000													
6. P fistulosus	0.030	0.034	0.027	0.000	0.000												
7. C colocynthis	0.027	0.032	0.025	0.002	0.002	0.002											
8. Sample	0.025	0.030	0.023	0.009	0.009	0.009	0.007										
9. C argyrosperma	0.025	0.030	0.023	0.009	0.009	0.009	0.007	0.000									
10. C palmata	0.023	0.027	0.020	0.007	0.007	0.007	0.004	0.002	0.002								
11. C foetidissima	0.023	0.027	0.020	0.007	0.007	0.007	0.004	0.002	0.002	0.000							
12. C maxima	0.023	0.027	0.020	0.007	0.007	0.007	0.004	0.002	0.002	0.000	0.000						
13. C.grandis	0.034	0.039	0.027	0.009	0.009	0.009	0.007	0.014	0.014	0.011	0.011	0.011					
14. C quinqueloba	0.032	0.037	0.025	0.007	0.007	0.007	0.004	0.011	0.011	0.009	0.009	0.009	0.002				
15. C adoensis	0.032	0.037	0.025	0.007	0.007	0.007	0.004	0.011	0.011	0.009	0.009	0.009	0.002	0.000			
16. C mackenii	0.032	0.037	0.025	0.007	0.007	0.007	0.004	0.011	0.011	0.009	0.009	0.009	0.002	0.000	0.000		
17. S melongena	0.108	0.119	0.103	0.108	0.108	0.108	0.108	0.110	0.110	0.108	0.108	0.108	0.108	0.105	0.105	0.105	

Figure 10. Estimates of evolutionary divergences between 16 input sequences and out-group using maximum composite likelihood method in MEGA4 software (pumpkin).

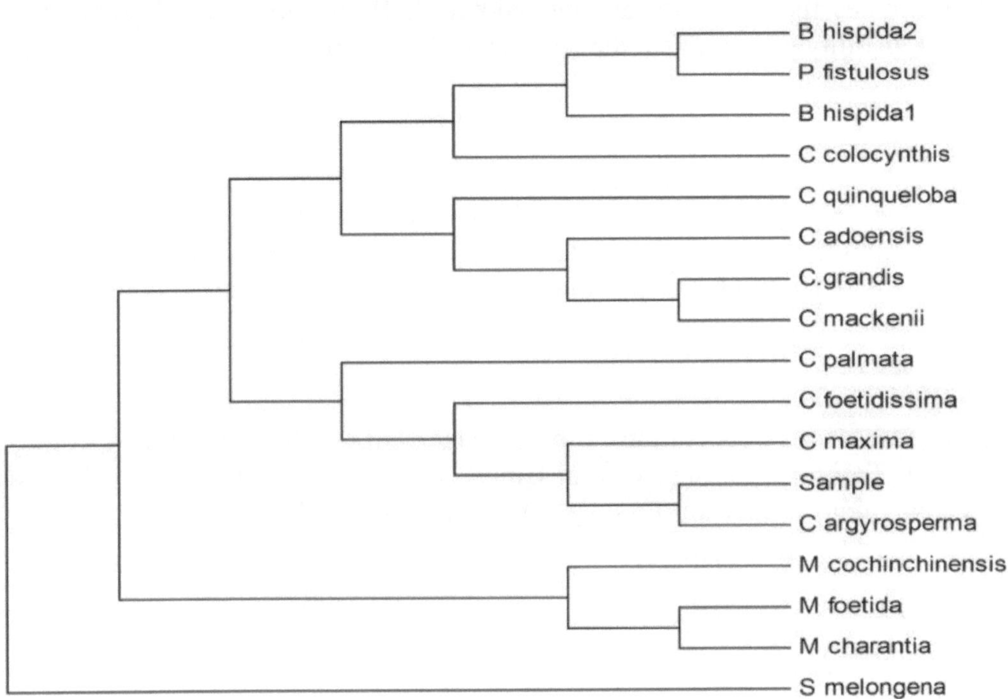

Figure 11. Evolutionary history of 17 taxa determined using neighbour-joining method (pumpkin).

4. Discussion

Phylogenetic nomenclature is a result of Darwins discovery that history of life can be best represented in tree-shaped diagrams. In the current era, the conventional phyolgenetic nomenclatures are supplemented with molecular biological tools which make the classification much more precise and accurate. In general the DNA or RNA sequencing gained superiority among evolutionary studies coupled with the support of Bioinformatics tools and software's.

Currently, the sequencing of entire DNA of an organism is long and expensive process. A typical molecular phyolgenetic analysis require 1000 base pairs. At any location within such sequences, the bases found in a given position may vary between organisms. The particular sequences found in a given organism is referred as its halotype. In principle, since there are four base types, with 1000 base pair, there could have 41000 distinct halotypes. In a particular species or in a group of related species, there would have few variations and most of them are co-related with few distinct haplotypes.

Once divergences between all pairs of samples were determined, statistical cluster analysis and dendrogams examines the similarity among halotypes. Bootstrapping and jackknifing further increase the reliability estimates for the position of haplotypes within the evolutionary tree.

5. Conclusions

The reliability of molecular phylogenesis can be affected by myriad, long branch attraction, saturation and taxon sampling problems. So the selection of the models plays a key step in the success analysis. More detailed work on the evolution within the Cucurbitaceae is needed for crop improvement.

Acknowledgements

The authors are grateful for the cooperation of the management of Mar Augsthinose college for necessary support. Technical assistance from Binoy A Mulanthra is also acknowledged. We also thank an anonymous farmer for the supply of plant materials for our study.

References

Badola HK, Aitken S (2010). Biological resources and poverty alleviation in the Indian Himalayas. *Biodiversity* **11**(3-4) 8-18.

Esteras C, Nuez F, Picó B, YiHong W, Behera TK, Kole C (2011). Genetic diversity studies in Cucurbits using molecular tools. *Genetics, Genomics and Breeding of Cucurbits* **2**(1) 140-198.

Gong L, Paris HS, Stift G, Pachner M, Vollmann J, Lelley T (2013). Genetic relationships and evolution in Cucurbita as viewed with simple sequence repeat polymorphisms: the centrality of C. okeechobeensis. *Genetic Resources and Crop Evolution* **2**(1) 1-16.

Robinson RW, Decker-Walters DS (1997). *Cucurbits.* Cab international. Wallingford.

Kocyan A, Zhang LB, Schaefer H, Renner SS (2007). A multi-locus chloroplast phylogeny for the Cucurbitaceae and its implications for character evolution and classification. *Molecular Phylogenetics and Evolution* **44**(2) 553-577.

Jeffrey C. (2005). A new system of Cucurbitaceae. *Bot. Zhurn, 90,* 332-335.

Lebeda A, Widrlechner MP, Staub J, Ezura H, Zalapa J, Krˇistkova E (2007). Cucurbits (Cucurbitaceae; Cucumis spp., Cucurbita spp., Citrullus spp.). In: Singh RJ (ed) Genetic resources, chromosome engineering, and crop improvement: vegetable crops. CRC Press, Boca Raton, pp 271–376

Gwanama C, Labuschagn, MT, Botha AM (2000). Analysis of genetic variation in Cucurbita moschata by random amplified polymorphic DNA (RAPD) markers. *Euphytica* **113**(1) 19-24.

Wu J, Chang Z, Wu Q, Zhan H, Xie S (2011). Molecular diversity of Chinese *Cucurbita moschata* germplasm collections detected by AFLP markers. *Scientia Horticulturae* **128**(1) 7-13.

Paris HS, Yonash N, Portnoy V, Mozes-Daube N, Tzuri G, Katzir N (2003). Assessment of genetic relationships in *Cucurbita pepo* (Cucurbitaceae) using DNA markers. *Theoretical and Applied Genetics* **106**(6) 971-978.

Gong L, Paris HS, Nee MH, Stift G, Pachner M, Vollmann J, Lelley T (2012). Genetic relationships and evolution in *Cucurbita pepo* (pumpkin, squash, gourd) as revealed by simple sequence repeat polymorphisms. *Theoretical and Applied Genetics* **124**(5) 875-891.

Wilson HD (1989). Discordant patterns of allozyme and morphological variation in Mexican Cucurbita. *Systematic Botany* **1**(2) 612-623.

Ruiz RAT, Hemleben V (1991). Use of ribosomal DNA spacer probes to distinguish cultivars of Cucurbita pepo L. and other Cucurbitaceae. *Euphytica* **53**(1) 11-17.

Baranek M, Stif, G, Vollmann J, Lelley T (2000). Genetic diversity within and between the species *Cucurbita pepo, C. moschata* and *C. maxima* as revealed by RAPD markers. *Report-Cucurbit Genetics Cooperative* **23**(1) 73-77.

Ferriol M, Pico MB, Nuez F (2003). Genetic diversity of some accessions of *Cucurbita maxima* from Spain using RAPD and SBAP markers.*Genetic Resources and Crop Evolution* **50**(3) 227-238.

Porebski S, Bailey LG, Baum BR (1997). Modification of a CTAB DNA extraction protocol for plants containing high polysaccharide and polyphenol components. *Plant Molecular Biology Reporter* **15**(1) 8-15.

Kidwell KK, Osborn TC (1992). Simple plant DNA isolation procedures. In *Plant genomes: methods for genetic and physical mapping* (pp. 1-13). Springer Netherlands.

Zhang LB, Renner S (2003). The deepest splits in Chloranthaceae as resolved by chloroplast sequences. *International Journal of Plant Sciences* **164**(S5) S383-S392.

Haas SA, Hild M, Wright AP, Hain T, Talibi D, Vingron M (2003). Genome scale design of PCR primers and long oligomers for DNA microarrays. *Nucleic acids Research* **31**(19) 5576-5581.

Rychlik WJSW, Spencer WJ, Rhoads RE (1990). Optimization of the annealing temperature for DNA amplification in vitro. *Nucleic acids Research* **18**(21) 6409-6412.

Meier A, Sander P, Bottger EC (1996). Protocol 4. Elimination of contaminating DNA within PCR reagents. PCR Essential Techniques, John Wiley & Sons, New York, 19.

Erlich HA (1989). *PCR technology. Principles and applications for DNA amplification.* Stockton press.

Saiki RK (1989). The design and optimization of the PCR. PCR technology: Principles and applications for DNA amplification, 7-16.

Thompson JD, Higgins DG, Gibson TJ (1994). CLUSTAL W: improving the sensitivity of progressive multiple sequence alignment through sequence weighting, position-specific gap penalties and weight matrix choice. *Nucleic acids Research* **22**(22) 4673-4680.

Chenna R, Sugawara H, Koike T, Lopez R, Gibson TJ, Higgins DG, Thompson JD (2003). Multiple sequence alignment with the Clustal series of programs. Nucleic acids research, 31(13), 3497-3500.

YOUR KNOWLEDGE HAS VALUE

- We will publish your bachelor's and
  master's thesis, essays and papers

- Your own eBook and book -
  sold worldwide in all relevant shops

- Earn money with each sale

Upload your text at www.GRIN.com
and publish for free